Number without a worksheet

for Reception and Year 1

Early Childhood Mathematics Group

Introduction

This book has been written by the Early Childhood Mathematics Group as a companion booklet to the ATM activity book *'Exploring mathematics with younger children'*.

The transition between R and Y1 is a challenge for both learners and teachers. Members of the ECMG have chosen their favourite number activities to help bridge the gap without resorting to worksheets. In *'Number without a worksheet'* you will find twelve core activities, each with extensions, simplifications and suggestions for developing the ideas, together with a list of resources and the identification of the mathematical learning. The activities give the help you need to save planning time and to encourage children to think mathematically.

The Early Childhood Mathematics Group is one of ATM's working groups.

The group started in 1996 and meets twice a term in London at the Institute of Education in Bedford Way, W1. The members of the group are practitioners working in the foundation stage, mathematics education lecturers, local education advisors, consultants, teachers and those interested in early years' education. At the meeting we discuss current early years' education issues as they relate to the teaching and learning of mathematics. We share problems and solutions and look at the realities of classroom practice. We also take on tasks such as writing responses to curriculum documents, writing articles for professional publications and developing curriculum materials.

With contributions from

Sheila Ebbutt

Cathy Murphy

Carole Skinner

Val Shaw

Pam Baldwin

Jenny Mitchell

Mary Southall

Siobhan Skeffington

Grace Cook

Jean Miller

Alison Millett

Sue Gifford

Design: DCG Design, Cambridge
Illustrations: Tamaris Taylor

Contents

		page
1	Count the Spaces	4
2	Flic Flacs (Eleflips)	6
3	How Many in the Bag?	10
4	Sea Shells	12
5	Dinosaur Adding	14
6	Vote For	16
7	Triangles	18
8	Under the Blanket	20
9	Line Up!	22
10	Counting Round	24
11	And One More	26
12	Line Game	28
	Photocopiable sheet 1	30
	Photocopiable sheet 2	31
	Photocopiable sheet 3	32
	Photocopiable sheet 4	33
	Photocopiable sheet 5	34

1 Count the Spaces
A game played in pairs

Mathematics learning

Understanding addition by combining sets to make a total

Complementary addition: how many more make 12

Predicting

Reasoning

Checking

Being systematic

Resources

1-6 dice

counting objects such as beads in a container

12 small yoghurt pots or two egg boxes for each pair of children

wooden numerals or cards

Main activity

Each child rolls the dice and collects that number of beads. They then take turns to put one bead in each compartment in the egg box until they have both used all their collected beads. The children then count how many compartments altogether have a bead in and how many compartments are empty. They then find the numeral card that matches the number of empty compartments. That is their score for that round.

The children return all their beads and continue to roll the dice once each. They count out the beads, put them in the compartments and count how many compartments are empty. They collect the appropriate numeral card, and keep it.

Play six rounds and discuss which numeral cards have been collected.

Extension

One child collects the addition total numeral and the other the complement to 12 numeral and they match them together.

Simplification

Use 1-3 dice and one egg box. Count how many altogether after two rolls of the dice, and collect that numeral.

Questions

- How did you know how many to take?
- How many spaces are there left to fill?
- What happened that time?
- What is the highest number you collected?
- What is the lowest number you collected?
- What do you think about this game?

Variation

Use a 1-5 spinner and play with real 1p coins and talk about about making 10p. Use bun tins and fill with playdough buns. Use Numicon and fill up with the pegs.

Outdoors

Group activity: Chalk a large grid on the floor and play with beanbags and large dice.

Recording suggestion

Provide cards with addition and equals signs and encourage the children to make number sentences or record using symbols and drawings to represent the results of the activity.

2 Flic Flacs (Eleflips)
A large-group activity

Mathematics learning

Number bonds to 10
Counting reliably to 10
Predicting
Reasoning
Checking
Being systematic

Resources

flic flac sheet for 10 (see photocopiable sheets 1 & 2)

Main activity

Show children a sheet of 10 flic flacs. Ask how many elephants they can see. Establish 10 by pointing and counting one by one.

Fold the paper along one fold and ask how many *they* can now see (for example, 8).

Ask how many elephants the *adult* can now see (for example, 2).

Check by counting on using fingers or on a number line.

Repeat with different folds.

back

front

Extension

Use flic flacs with multiples, such as two wellington boots or five fingers in each 'cell'.

Flic Flacs
(continued)

Simplification

Use flic flacs for number bonds to 5

Questions

- How can we be sure?
- How do you know?
- How did you work it out?
- How could you work this out?
- If there were only eight elephants and I could see three, how many could you see?
- How many more would we need to make …?
- What if … I could see two elephants?

Variation

Independent activity:
A group with a group 'leader'. Sticks of 10 interlocking cubes. Break some off, how many now?

Outdoors

Group activity: Ask 10 children to be the elephants. Use a screen for some to hide behind. Children close eyes and open them again. Ask how many in front and behind the screen.

Use beanbags on a large grid and cover some with a sheet.

Recording suggestion

Encourage children's own recording of the different ways of making 10; 1 and 9, 2 and 8 … Adult models an alternative method of recording.

3 How Many in the Bag?
An independent activity

Mathematics learning

Counting reliably to 10

Estimating to 10

Recording using numerals

Predicting

Reasoning

Checking

Being systematic

Resources

pencil cases, make up bags, sponge bags

interesting counting objects

sticky notes

pens

Main activity

Children choose some objects to hide in the bags. They swap bags and estimate how many there are inside. They record their estimate on a sticky note and stick it to the bag. They each count to check, and correct the sticky note if necessary.

Extension

Ask the children to say how many there would be if two more objects were added or taken away.

Use smaller objects and larger numbers.

Simplification

Hide a small number of objects in a bag. Pass it round the group and encourage the children to feel the bags and estimate how many.

Questions

- Shake the bag. How many do you think are inside?
- Why do you think that?
- Were you close?
- What could you write to remind you how many objects are in the bag?
- How many does it say are inside this bag?

Variation

Repeat main activity using real 1p coins and a purse.

Invite children to fit as many objects in the bag as they can.

Variation for outdoor area

Use carrier bags and groceries for children to estimate how many. Put objects in a bag high up on the wall and ask the children to guess how many apples are in the bag today. How many do you think? Is that more or fewer apples than yesterday?

Recording suggestion

Order the bags from most to least. Display numeral cards and bags/boxes of objects to match up.

4 Sea Shells
A game played in pairs

Mathematics learning

Counting reliably to 5 or 10

Number bonds to 5 or 10

Predicting

Reasoning

Checking

Being systematic

Resources

real shells

base boards (see photocopiable sheet 3)

Main activity

Children work in pairs. Each child has a base board and five shells. The children are asked to place their five shells on their board in any way they like.

Children take it in turns to find a different way to place the shells for each other

Ask: Can we find all the different ways of arranging five shells in the sand and the sea?

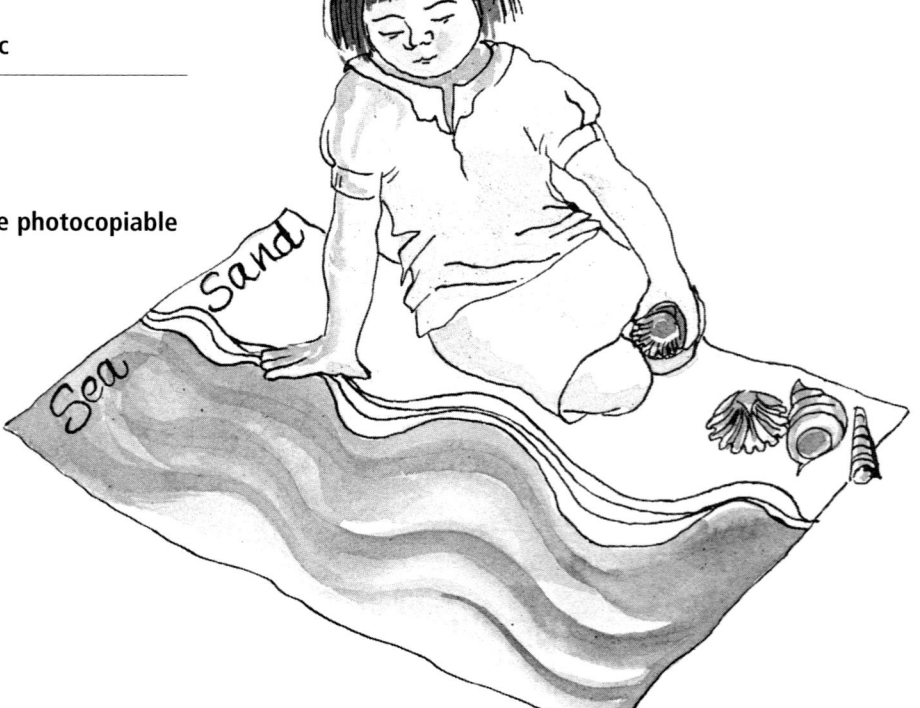

Extension

Repeat the activity with a different number of shells. What do you notice?

Predict how different many ways there would be with seven or eight shells.

Simplification

Use fewer shells and ask the children each time to describe how they have placed their shells.

Questions

- Have we found all the ways?
- How do we know?
- Are three shells in the sea and two on the sand different from two shells in the sea and three shells on the sand?
- Can we have all the shells on the sand?
- Tell me about this way of arranging your shells
- What if... we played with seven shells?

Variation

There are five shells altogether, four are on the sand how many are under the sea?

Outdoors

Use other materials; for example, five balls in and outside a hoop or bucket of water.

Recording suggestion

Encourage children's own drawing. Place numerals on a blank sand/sea recording sheet.

5 Dinosaur Adding
A group activity

Mathematics learning

Count reliably to 10
Adding one more
Predicting
Reasoning
Checking

Resources

large cardboard box
A4 card and pens
Blutac
plastic counting dinosaurs
a wheeled truck

Main activity

Teacher and children together count one or more dinosaurs into the truck before it goes into the box. The truck is pulled into the box and a dinosaur is dropped into the truck. The children predict how many dinosaurs will now be in the truck before it comes out.

Extension

What if... you knew one more dropped in and you knew how many came out. How could you work out how many went in to start with?

Simplification

Work within five. Children have five dinosaur counters to model what happens.

Questions

- How many dinosaurs do you think will come out of the machine?
- How did you work it out?
- What do you think?
- What if five dinosaurs were put in to start with? ...12 dinosaurs? ...six dinosaurs?

Variation

With pairs of children; roll dice or turn a card from 0 to 10 or 20 to say how many objects go in to begin. Second child predicts what will happen.

Outdoors

Use a large box for a child to fit inside and be the machine operator.

Recording suggestions

Encourage children's own methods, 'make a note to show what happens to the number of dinosaurs each time'.

6 Vote For
A group activity

Mathematics learning

Counting and numeral recognition to 20

Data handling

Predicting

Reasoning

Checking

Being systematic

Resources

name card or picture/photo card for each child

Blutac

two columns drawn on large sheet of card to fit children's cards

numeral cards

A4 card and pen (for writing the daily question)

Main activity

Have a daily question such as:

Which of these . . . two stories, two songs, two jigsaws, two games . . . shall we have today?

Children 'vote' for one or the other by placing their card on the bar chart.

Count each choice and find the numeral to match. Read, sing or play the voters' choice!

Extension

Extend the vote to between three choices:

What shall we have as a role play area?

What is your favourite fruit?

What flavour cookies shall we cook tomorrow?

Simplification

Provide illustrations of items to vote with rather than using children's names.

Questions

- Which was most popular today?
- What does our chart tell us today?
- Before you count, estimate how many voted for ... provide silly suggestions like 50? 2? three and a half?
- What is the difference?
- Was the vote close today?
- What shall we vote on tomorrow?

Variation

Independent activity: Give children clipboards and encourage them to hold their own survey.

Outdoors

What apparatus shall we put out today?

Recording suggestion

Use an early years ICT package to produce a block graph of the vote. Encourage the children to vote 'on line'.

Ask a child to write up the result of today's vote for the parents' noticeboard.

7 Triangles
An independent activity

Mathematics learning

Counting reliably to 20

Adding three small numbers to 20

Recognising numerals to 20

Checking

Being systematic

Resources

laminated triangles with a circle at each corner (see photocopiable sheet 4)

1-6 spotty dice

counting objects

wooden numerals or numeral cards

Main activity

Roll a dice. Collect that number of objects. Place in a circle. Repeat twice for remaining circles. Count how many altogether. Find that numeral and place in centre.

What totals can you make?

Extension

Place a number in the middle. How many ways are there to arrange that number of counters into the three circles?

Simplification

Use a 1-3 dice.

Questions

- How did you work it out?
- What else could it be?
- What couldn't it be?
- Cover one group with a sheet. How many are under the sheet?

Variation

Use a different shape, such as a square or a pentagon.

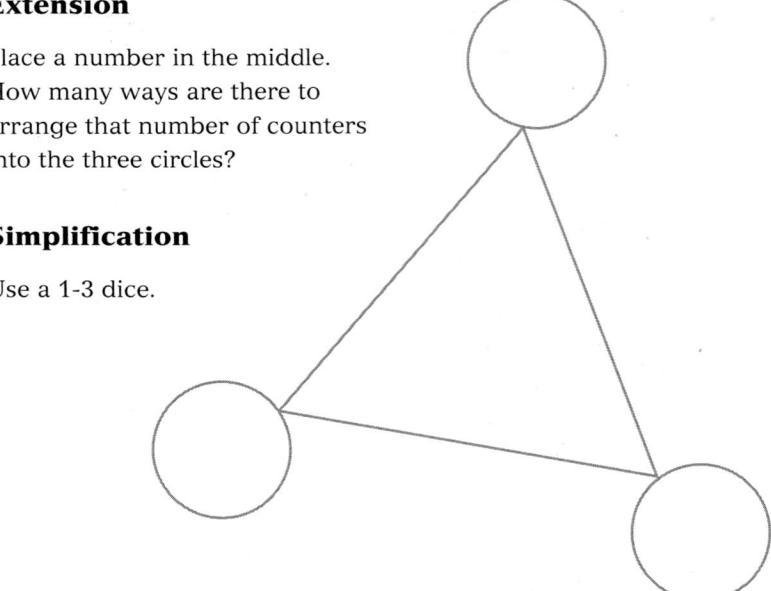

Outdoors

A group game played with two hoops or chalked circles and a 1-6 dice. Toss the dice and that many children stand in the first hoop. Toss the dice twice more and each time that many children stand in a hoop. Count how many children altogether.

Recording suggestion

'Find a way of recording the totals you make.'

Give each child a number track to 20. They colour the totals they make.

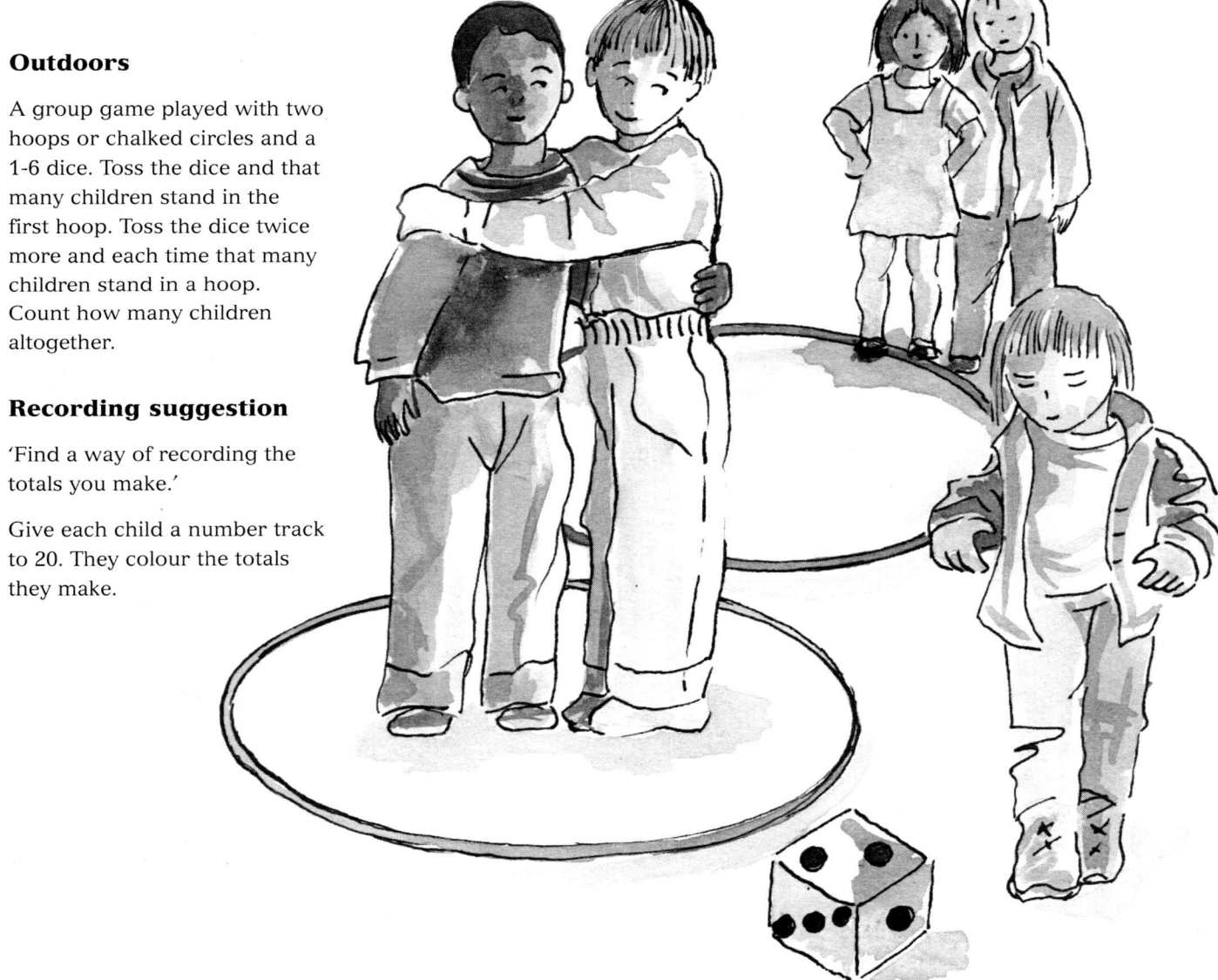

8 Under the Blanket
A group activity

Mathematics learning

Number bonds to 10
Calculating within 10
Predicting
Reasoning
Checking
Being systematic

Resources

piece of cloth for the 'blanket'
soft toys

Main activity

Each session, choose a number appropriate to your class or group that you want to reinforce the bonds for. Show the children this many soft toys and all count them. Cover them all with the blanket. Remove some and place them next to the blanket.

How many are under the blanket?

After this has been modelled a few times one of the children takes over.

Extension

Work with a larger number or roll a 1-6 dice each time to determine the number to work with (this is more of a challenge!)

Simplification

Build up number bonds over time with totals 3 to 6. Children try out each 'hiding' using a handkerchief and small counting toys.

Questions

- How do you know?
- Are you sure?
- How could we check?
- Have we done this one before?
- If I took out another how many would there be?

Outdoors

Place a given number of toys in a sand tray. Cover with sand. Remove some.

Recording suggestions

Children record the number they think are under the blanket on whiteboards each time.

Variation

Use a cup to cover transparent counters on the OHP.

9 Line Up!
A group activity

Mathematics learning

Counting up to 20 and above

Predicting

Checking

Being systematic

Resources

1-minute timer

teddy bear counters

Main activity

Sit in a circle each holding a bear counter. Chalk a long line on the floor. Start the timer. One at a time, around the circle, each child places a bear in the line and the whole group counts aloud. Stop when the 1-minute timer runs out. Ask 'How many bears are in the line this time?' The whole group say the number aloud.

Extension

Place a few bears on the line before starting (counting on).

Simplification

Use a shorter timer so smaller amounts are counted out.

Questions

- How many is that so far?
- Can you say how many are in the line without counting again?
- Is that more or less (fewer) than last time?
- What would the next number have been?
- What do you think might happen this (next) time?'
- How many do you think . . . ?

Variation

Work in pairs and take it in turns to thread beads on a lace. Count how many beads were threaded in a minute.

Outdoors

Outdoor challenges could include:

Taking it in turns to remove a potato from the bucket, 'How many can we take out in a minute?'

How many rings can we put onto a post in one minute?

Recording suggestions

Children record the numbers counted in one minute on whiteboards or a flipchart.

10 Counting Round
A group activity

Mathematics learning

Sequencing numbers to 12

Counting on and back to 12

Using days of the week and months of the year

Reading the numbers on a clock

Predicting

Reasoning

Representing

Resources

1 to 12 numbered carpet tiles

cards showing the months of the year, the days of the week

Main activity

Arrange 12 carpet tiles in a circle like a clock face. Go through the months of the year with the children, and agree the number of each month.

Children sit next to the number of the month that corresponds to their birthday month (you may need to remind children which month their birthday is in). If there are more than 12 children, they sit in line behind the number. Count the number of birthdays for each month.

Establish the current month. Invite children to step from that month to their own birthday month to find out how many months it is until their birthday.

Extension

Shuffle the cards showing the names of the months. Children take turns to pick a card, find the number that represents that month, and swap with the child who is sitting there by counting round and finding how many months away they are. Allocate months to seasons.

Simplification

Match the months to the numbers. Step one month forward and one month backwards. Discuss what happens when you get to December.

Questions

- How many steps do you think you'll take to get from 7 to 10?
- If we count on 2 and finish at 6, which number do we start at?
- Are there more birthdays in March or in April?
- If your birthday is next month, what number month will that be?

Variation

Use the numbers 1-7 in a circle to represent the days of the week, and use word cards showing the names of the days. Children count how many 'sleeps' it is till, for example, Saturday.

Use the numbers as the clock face, and talk about significant times during the day, and how many hours different sessions of the day take.

Outdoor area

Use a chalked 1-12 circular number track, or use the numbered tiles, and have the cards showing the months available. Have a giant spinner in the centre of the circle and a large 1-6 dice. Spin the spinner to show the starting month, roll the dice, and move clockwise that number of steps. Everybody chants together the months in order until they get to the month the child has landed on.

Recording suggestion

Make a circular representation of their birthday months using the children's names or photographs.

11 And One More
A group activity

Mathematics learning

Counting to 20
Adding one more
Predicting
Reasoning
Checking
Being systematic

Resources

'one more' strips (see photocopiable sheet 5).

counting objects such as counters, bricks, mini-dinosaurs, teddies, and so on

Main activity

Place one object on the first circle. Place one more than this on the second circle, and one more on the next circle and so on.

Ask the children to say what you are doing.

Repeat with a different number of objects in the starting circle.

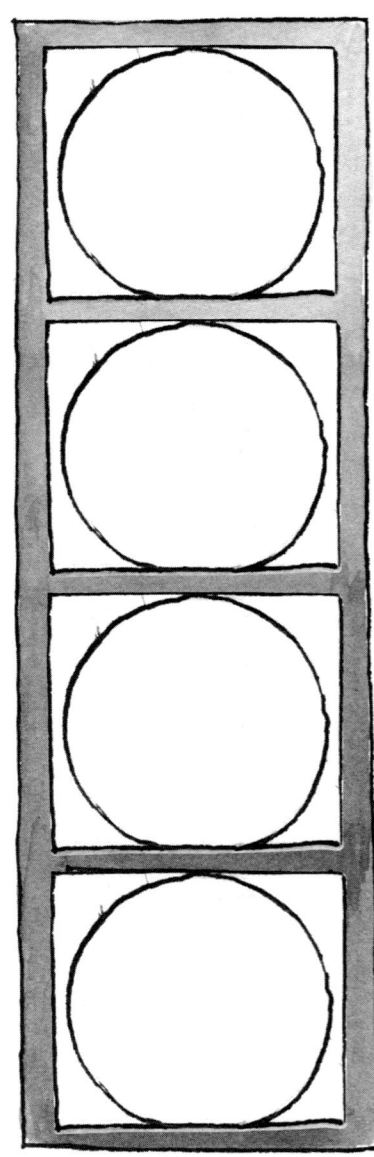

Extension

Place objects on the third or second circle to fill backwards and forwards.

Use a longer strip and use numerals.

Simplification

Use two plates and some biscuits. However many biscuits are put on the first plate the children must put the same number and one more on the second plate.

Questions

- What else could we use ... try?
- What did you notice? Tell me what you see.
- How are the numbers growing?
- What would be in the next space (not shown)?
- What if we ...

Variation

Build staircases using interlocking cubes or number rods.

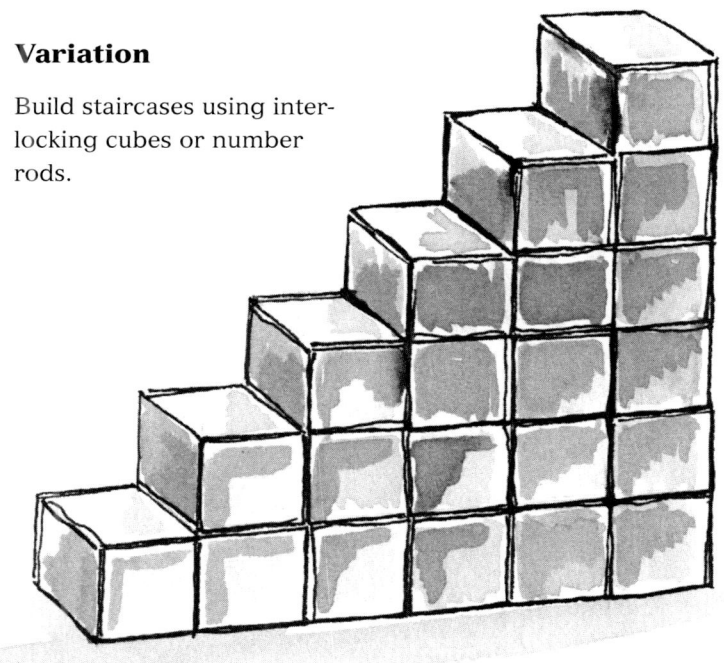

Outdoors

A large group activity. Use chalk circles and children. Draw four large circles in a line. One child rolls a large dice to set the first number. This number of children fit into the first circle, and the remaining children fill the rest, making each circle one more than the previous one.

Recording suggestion

Encourage child's individual recording of number in each circle.

12 Line Game
A game played in pairs

Mathematics learning

Counting to 30
Recognising numbers to 30
Predicting
Checking

Resources

0-30 table top number line
1-6 dice
two counters
washable pen
small toys

Main activity

Pairs take turns first of all to ring three numbers each anywhere on the number line. They then take turns to roll the dice and move their own counter that number of markers along the line. If they land on any circled number they win a toy. When both players have reached the end of the line, the player who has won the most toys wins.

Extension

Use a 0-50 line.

Use a 0-100 number line, and circle 10 numbers each; start with 10 counters and they lose one each time they land on a circled number.

Simplification

Use a 0-20 line.

Use a 0-10 line and a 1-3 dice.

Questions

- What do you need to roll to get to ...?
- How many jumps is it to the next circled number?
- How many toys did you win?
- Which circled numbers didn't you land on?
- Did you finish exactly on 30 or did you go past it?

Variation

One child starts at 30 and goes back to 0, while the other starts at 0.

Cover the numbers on the line with sticky notes or counters. Use number cards 0-30, shuffled in a pile. Children take a card in turn and decide where that number is on the line. They remove the note or counter and keep it if they are correct.

Outdoors

Chalk a long line outside and provide large foam dice and beanbags. Children could chalk their own line to play on.

Recording suggestion

Encourage children to write the numbers they landed on and to compare these with their partner's.

Photocopiable sheet 1 (Activity 2)

Photocopiable sheet 2 (Activity 2)

Photocopiable sheet 3 (Activity 4)

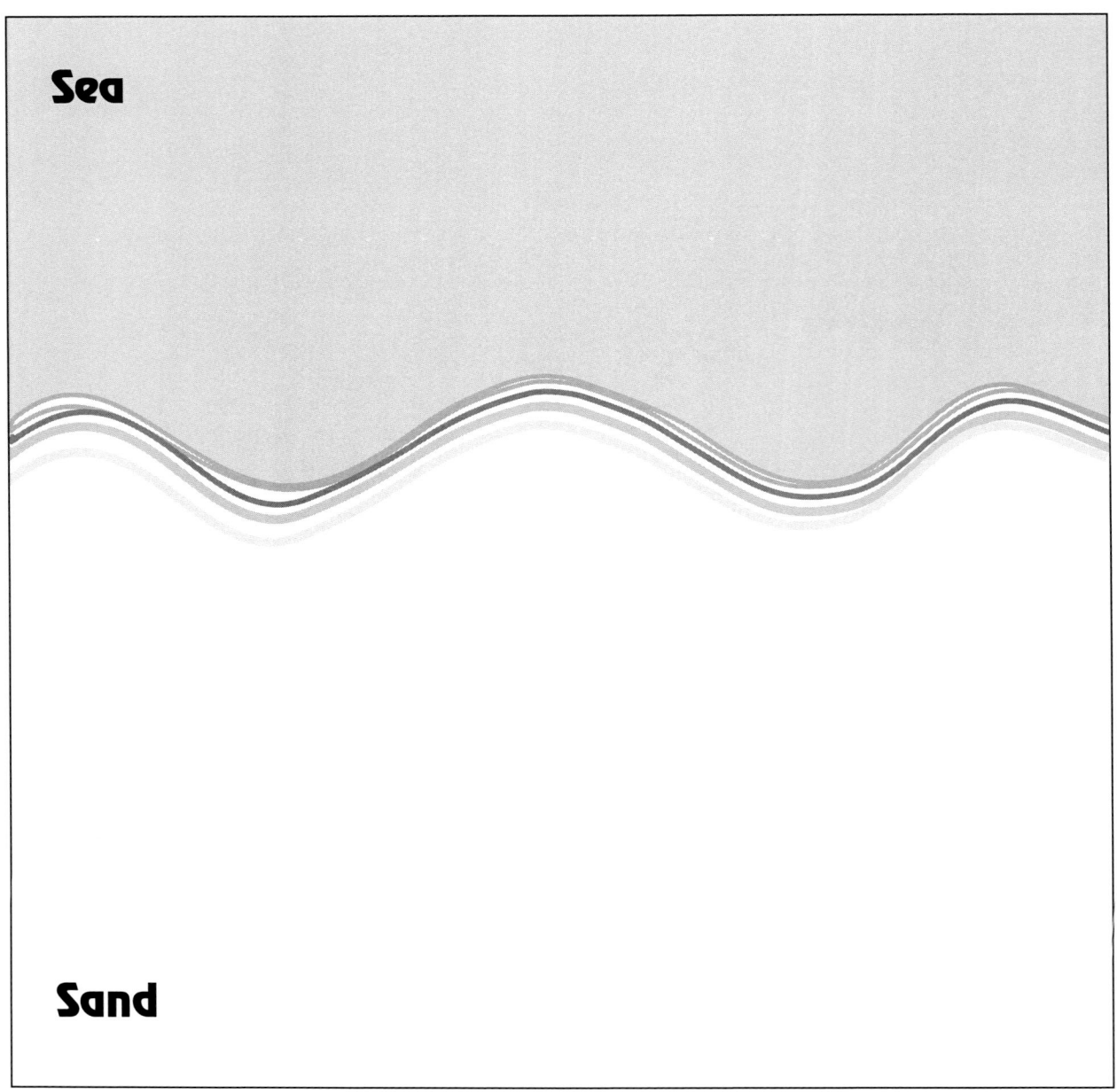

Photocopiable sheet 4 (Activity 7)

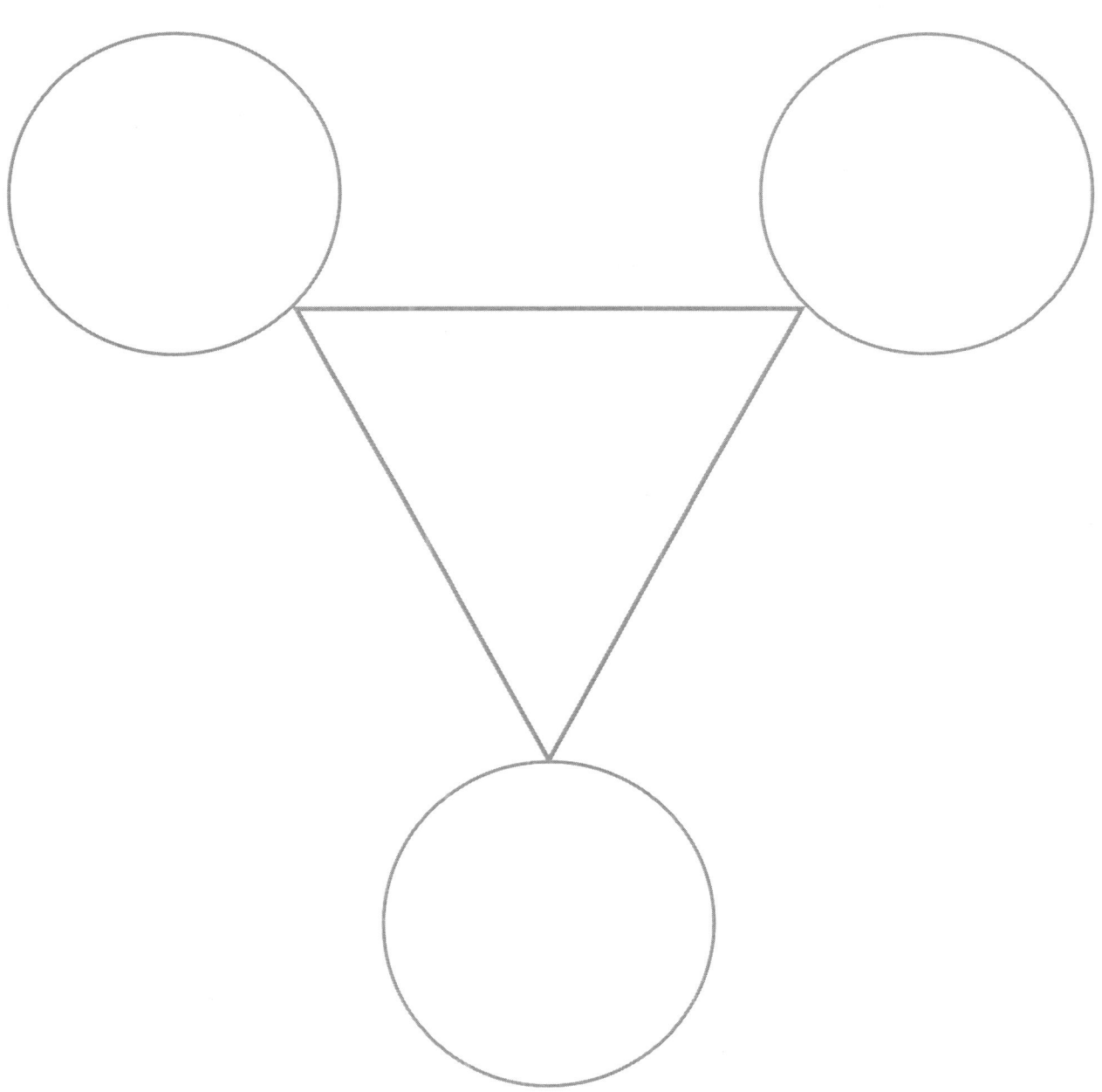

Photocopiable sheet 5 (Activity 11)

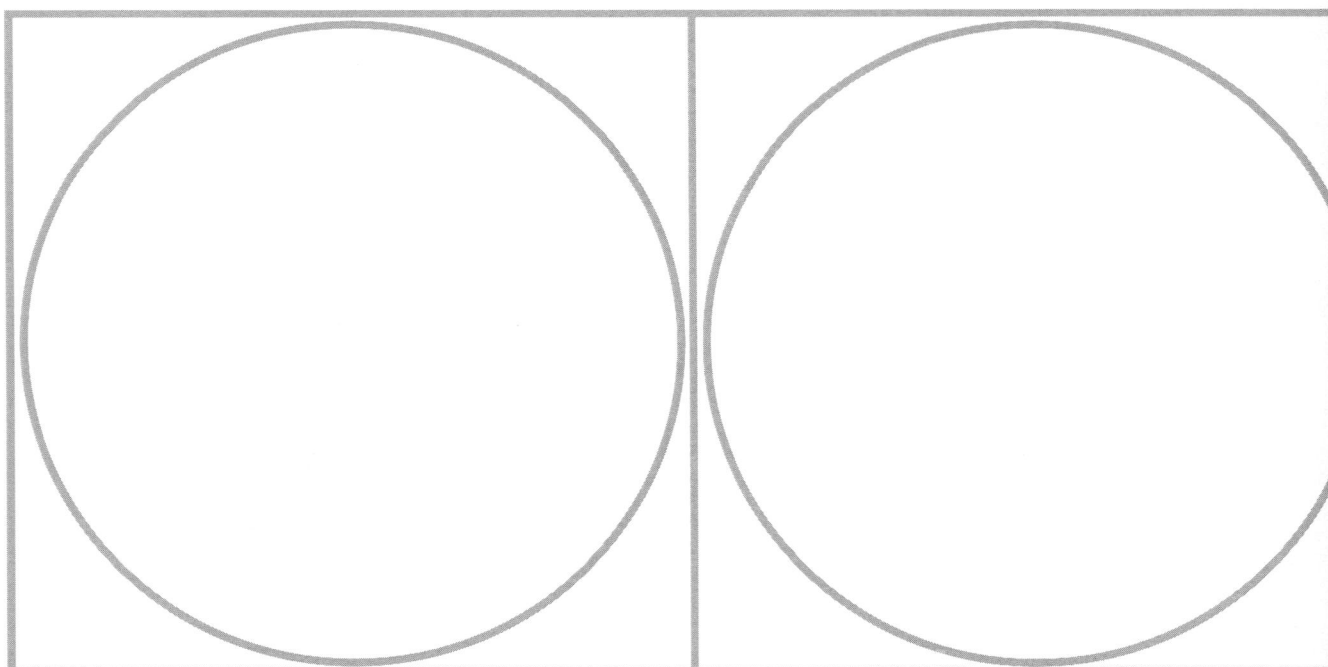

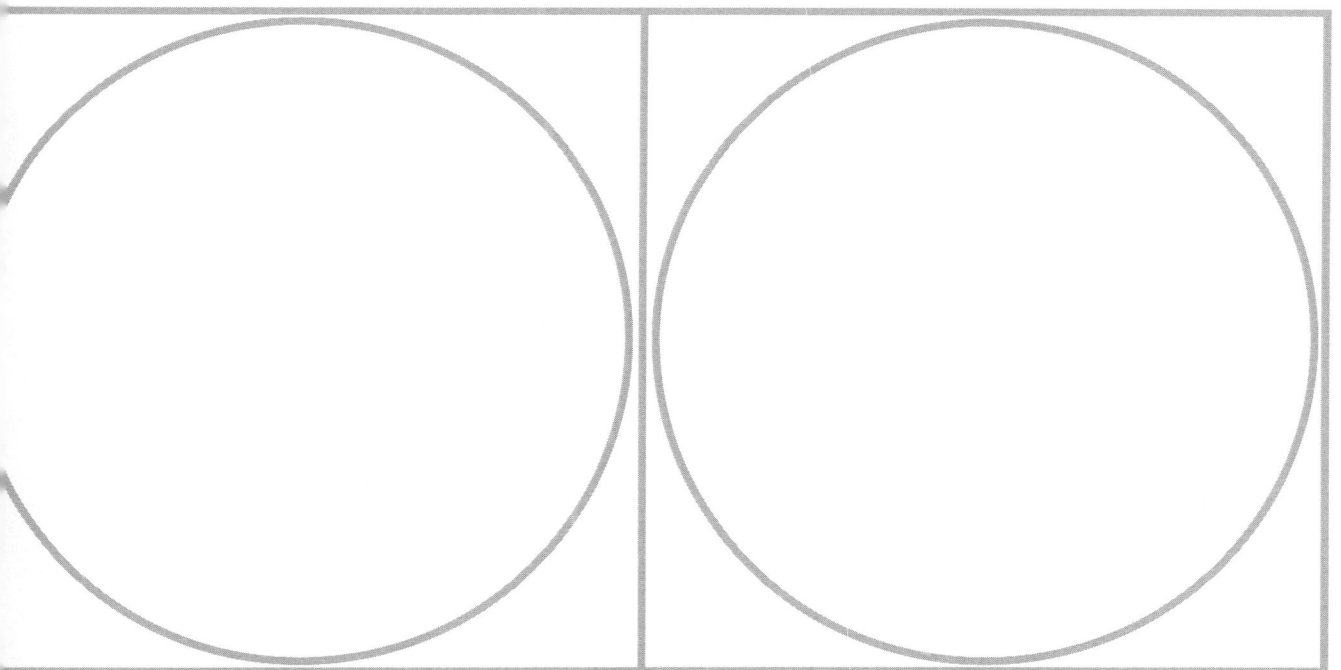

36